AB UNSIMPLIFIED

If a person was to explain what A M*ultiplied* by B is they might say "those letters may be any number, e.g. A may be the quantity of identical ladybirds, whereas B may be the quantity of eyes that one ladybird has, and the answer to AB would be the total number of eyes that someone could count."

Another way *Multiplication* could be explained is by picking a handful of balls from a table and counting how many balls fit into your hand. You can then do *Multiplication* to find how many balls you will have when a specific number of handfuls are taken.

The Oxford dictionary just defines *Multiplication* as adding a number to itself a specified number of times so the sum 4 x 8 may be thought of as 4 lots of 8 or 8+8+8+8 but the way you go about adding numbers together is entirely up to you. There is no specified method to *Multiply* numbers.

Most people who grasp the concept of *Multiplication* are happy to accept that A x B or AB equals C, and only learn the most basic way to turn *Multiplication* problems into addition problems to some extent.

E.g. if A = 3, A can be broken down into 1 & 2

3 x B = (1 x B) + (2 x B)

To ensure A is split into two numbers which equal A when added together, the below equation can be used:

A x B = (A − N) x B) + (N x B)

Both A and B may be divided any number of times before being multiplied, and then added to give the answer C.

This understanding of *Multiplication* helps to create methods for doing *Multiplication*. I have made one whereby *Multiplications* are solved only via adding together numbers from the 1,5,and 10 times tables.

I give four coloured sticks a value:

STICK COLOUR	INDIVIDUAL VALUE
Black	-1
White	+1
Red	+5
Green	+10

To solve 7 x 6, you must first make each value using the sticks provided. 7 is made by using one red stick worth 5, and two white sticks worth 2. Finally 6 is made by using one red stick and one white stick.

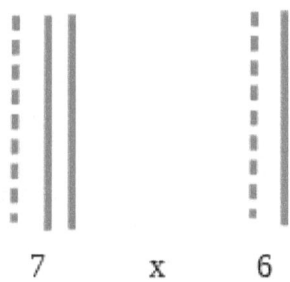

7   x   6

It is a little like writing Roman Numerals

To solve this *Multiplication* - overlap the sticks, and where they cross multiply the value of each stick by the value of the other, and then add them together.

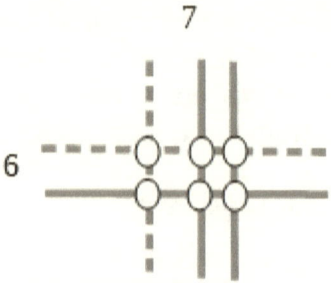

25 + 5 + 5 + 5 + 1 + 1 = 42

The limited number of combinations makes problems like these very easy to solve and the use of the black stick can reduce the number of sticks required.

Intersecting Values:

White & White	White & Red	White & Green	Red & Red	Red & Green	Green & Green
1 x 1	1 x 5	1 x 10	5 x 5	5 x 10	10 x 10
1	5	10	25	50	100

1x	Red & Red	:	25
3x	White & Red	:	15
2x	White & White	:	2
			42

I deliberately did not show the answer when a black stick overlaps another stick because it can be worked out! Try to do A x B using zero, one, or more black sticks, and you will find that a minus number times another minus number must be a positive number.

A good understanding of *Multiplication* is needed to unsimplify AB. No doubt there will be people who think AB only equals AB, and think I'll show this:

A x B = (A x B) - 4) + 4) / 7) x 7)

Such an equation may even be squared provided the equation is square rooted so that everything cancels revealing that AB = AB i.e. itself.

However, what AB does equal is more astonishing than this due to patterns in the Multiplication grid:

x	1	2	3	4	5	6	7	8	9	10	11	12
1	1	2	3	4	5	6	7	8	9	10	11	12
2	2	4	6	8	10	12	14	16	18	20	22	24
3	3	6	9	12	15	18	21	24	27	30	33	36
4	4	8	12	16	20	24	28	32	36	40	44	48
5	5	10	15	20	25	30	35	40	45	50	55	60
6	6	12	18	24	30	36	42	48	54	60	66	72
7	7	14	21	28	35	42	49	56	63	70	77	84
8	8	16	24	32	40	48	56	64	72	80	88	96
9	9	18	27	36	45	54	63	72	81	90	99	108
10	10	20	30	40	50	60	70	80	90	100	110	120
11	11	22	33	44	55	66	77	88	99	110	121	132
12	12	24	36	48	60	72	84	96	108	120	132	144

I was taught to look at the rows and columns where numbers can be seen increasing in fixed increments but if you look at the numbers diagonally from any square number; it decreases by 1, then 3, then 5 and then the next odd number, and so on.

The amount subtracted from the square number is a square number. So if you couldn't remember what 7^2 equals you could subtract 1 from one 7 and add 1 to the other 7, multiply these numbers, and then add the first square number to it. If you forgot what 6 x 8 equals, the step can be repeated ensuring you add the next square number, and so on, see below:

7 x 7 = (6 x 8) + 1^2
7 x 7 = (5 x 9) + 2^2
7 x 7 = (4 x 10) + 3^2
7 x 7 = (3 x 11) + 4^2

This is expressed more clearly by the equation:

A x A = ((A-N) x (A+N)) + N^2

Writing an equation for non square numbers is a bit trickier. I started searching for an equation for Odd numbers multiplied by other Odd numbers, and Even numbers multiplied by other Even numbers because square numbers cannot be made via multiplying any Odd number by an Even number. But I had already found an equation for square numbers so I thought I could easily do this myself.

I made a list of multiplications.

Can you spot the pattern?

1 x 3 = 3
2 x 4 = 8
3 x 5 = 15
4 x 6 = 24
5 x 7 = 35
6 x 8 = 46
7 x 9 = 63
8 x 10 = 80
9 x 11 = 99
10 x 12 = 120

The answer is one less than the square of the average of A x B. There's an equation I had to remember in school to solve the nth term of sequences like this:

a + (n-1)d + ½ (n-1)(n-2)c

However, this doesn't tell you what AB equals.

Do you know what A x B equals in this sequence?

1 x 5 = 5
2 x 6 = 12
3 x 7 = 21
4 x 8 = 32
5 x 9 = 45

If you square the average of A x B you must subtract four this time to arrive at the answer. A larger square number must always be subtracted if there's a larger difference between A and B.

I found that when A and B are both Even or are both Odd then the following equation can be used:

$((B + A)/2)^2 - ((B - A)/2)^2$

When multiplying an Odd and Even number you can use the above equation and subtract one from A or B and depending upon which you must subtract either A or B to arrive at the correct answer.

If you look at 6 x 7 = 42

You could arrive at 42 by adding 5 lots of 7 and then adding one more 7. Or, you can add 6 lots of 6 and then add another 6.

To put this more simply any Odd number multiplied by an Even numbers where B is higher than A is:

$((B + A - 1)/2)^2 - ((B - A - 1)/2)^2 + A$

An Odd number *multiplied* by any whole number is always equal to an Even number.

If you look at an area of 3 x 4 and you double one number either 3 or the 4; the area will double but if you double both numbers then the area will be 6 x 8 i.e. quadruple the size of 12.

The following equation proves this:

A x B = (2A x 2B) / 4

To get a better picture of what happens I have shown some dots below:

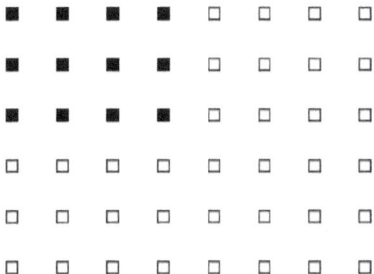

If I want to know what 3 x 4 is, and 3 x 5 using the same equation, I could make every number Even by doubling them and when I find the answers to 6 x 8, and 6 x 10 I can just divide them by 4.

But I do not have to multiply both numbers by 2 and divide by 4, provided the below equation is satisfied:

A x B = ((N x A) x (N x B)) / N^2

I can now make an equation for AB, see below:

A x B = ((B +A)^2 - (B - A)^2) / 4

I could unsimplify this equation forever because:

(B + A) = 1 + (A x B) − ((A − 1) x (B − 1))

However because the answer contains (A x B) I don't consider this a valid form of unsimplification.

There is a part that can be unsimplified though

$(B - A) = (B + A) - (A + A)$

$(B - A) = (B + B) - (B + A)$

If I wanted I can change $(B + B)$ into 2B, and place another number N into the equation to find $(B - A)$:

$(N \times (2B)) - (N \times (B + A)) + (N \times ((2B) - (B + A)))$

Instead of squaring the above to find $(B - A)^2$ I could use the unsimplified equation below:

$(B - A)^2 = (B^2 + A^2) \times 2 - (B + A)^2$

I could unsimplify $(B + A)^2$ in the following way:

$(B + A)^2 = (B^2 + A^2) \times 2 - (B - A)^2$

Using all of the above knowledge it is possible to usimplify AB to make a long equation:

$AB = ((((B^2 + A^2) \times 2) - (B - A)^2 - (((B^2 + A^2) \times 2) - (B+A)^2)) / 4$

I can quite easily make this equation longer still if I include "N" to the equation but there comes a point when an unsimplifed equation is unpractical.

Unsimplifying equations is undoubtedly a skill and I don't believe that it is taught despite it being a very good way to help improve logical thinking, and find different or perhaps better equations for AB.

I can show that $(B + A)^2 = (2AB) + B^2 + A^2$

	B	+	A
B	B^2		AB
+			
A	AB		A^2

I know that all 4 squares equal $(B + A)^2$ and I know subtracting both A^2 and B^2 leaves just 2AB, thus:

$2AB = (B + A)^2 - B^2 - A^2$

This may be rewritten as:

$2AB = (B + A)^2 - (B^2 + A^2)$

So...

$AB = ((B + A)^2 - (B^2 + A^2)) / 2$

Dividing by 2 is much easier than dividing by 4 but you could easily divide by any number provided you also multiply by half that value, see below:

$AB = (((B + A)^2 - (B^2 + A^2)) \times N) / 2N$

There are many other ways AB can be unsimplified and the next part of the book will look at how some equations which may seem entirely separate from AB but are in fact a part of AB. Such equations solve the nth term of number sequences.

I have made a grid below. At first glance it may look very much like the multiplication grid but by looking closely at the diagonal pattern you'll notice numbers get larger instead of smaller the further they are from the black squares. This is because on this grid AB is equal to $((B^2 + A^2) \times 2)$.

The other major difference is found by looking at the rows and columns; numbers increase forever and by greater amounts each step, and each row and column share the same sequence.

x	1	2	3	4	5	6	7	8	9	10	11	12
1	4	10	20	34	52	74	100	130	164	202	244	290
2	10	16	26	40	58	80	106	136	170	208	250	296
3	20	26	36	50	68	90	116	146	180	218	260	306
4	34	40	50	64	82	104	130	160	194	232	274	320
5	52	58	68	82	100	122	148	178	212	250	292	338
6	74	80	90	104	122	144	170	200	234	272	314	360
7	100	106	116	130	148	170	196	226	260	298	340	386
8	130	136	146	160	178	200	226	256	290	328	370	416
9	164	170	180	194	212	234	260	290	324	362	404	450
10	202	208	218	232	250	272	298	328	362	400	442	488
11	244	250	260	274	292	314	340	370	404	442	484	530
12	290	296	306	320	338	360	386	416	450	488	530	576

I can use these differences to insert an equation that I learnt at school, and show it is merely part of AB.

If you can't remember; a + (n-1)d + ½ (n-1)(n-2)c

a = the first term
d = the first difference
c = the change between one difference and the next

To find a grid pattern the equation can be used twice but this isn't quite straight forward.

The sequence goes like this: 4, 10, 20, 34, 52…

The first term = 4
The first difference = 6
The change between one difference and the next = 4

So part of the equation will look like this:

4 + (B – 1) x 6 + 0.5 x (B -1) x (B – 2) x 4

If I want to find the number where 7 x 3 is I can find the third number in the sequence, and then I add six not seven numbers to 20 i.e. 6 + 10 + 14 + 18 + 22 + 26 to get 116.

So the second sequence is a bit different to the first:

The first term = 0
The first difference = 6
The change between one difference and the next = 4

This part of the equation will be:

(B – 1) x 6 + 0.5 x (B -1) x (B – 2) x 4

If I make 6 the first term the equation is:

6 + (B − 1) x 6 + 0.5 x (B -2) x (B − 3) x 4

Now I have found that $((B^2 + A^2) \times 2)$ is equal to:

(4 + (B − 1) x 6 + 0.5 x (B - 1) x (B - 2) x 4) + ((B - 1) x 6 + 0.5 x (B - 1) x (B - 2) x 4)

I can further unsimplify the below equation:

AB = $((((B^2 + A^2) \times 2) - (B - A)^2 - (((B^2 + A^2) \times 2) - (B+A)^2))/4$

by replacing $((B^2 + A^2) \times 2)$ with what I found.

I used to think that AB was a just a very small part of Mathematics, a branch if you like but everything you can learn about numbers, quantities and space can be shown to be merely a part of AB.

Further evidence is found by exploring the polygonal numbers including the ones below:

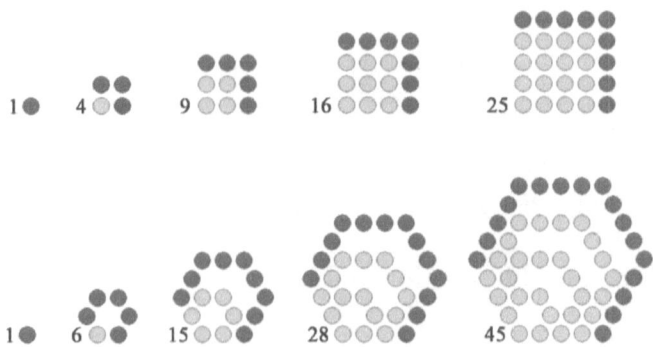

Here are some nth terms of polygonal numbers:

n	s										
1	1	1	1	1	1	1	1	1	1	1	1
2	2	3	4	5	6	7	8	9	10	11	12
3	3	6	9	12	15	18	21	24	27	30	33
4	4	10	16	22	28	34	40	46	52	58	64
5	5	15	25	35	45	55	65	75	85	95	105
6	6	21	36	51	66	81	96	111	126	141	156
7	7	28	49	70	91	112	133	154	175	196	217
8	8	36	64	92	120	148	176	204	232	260	288
9	9	45	81	117	153	189	225	261	297	333	369
10	10	55	100	145	190	235	280	325	370	415	460
11	11	66	121	176	231	286	341	396	451	506	561
12	12	78	144	210	276	342	408	474	540	606	672
13	13	91	169	247	325	403	481	559	637	715	793

If you look at each column from left to right, and the change between one difference and the next you will notice it increases by one each time.

If you look at each row, the numbers increase from left to right by a fixed number like the Multiplication grid except it increases by a Triangle number each time where the sequence goes 1, 3, 6, 10, 15, 21, 28 etc.

I can find the nth Triangle number by drawing them; adding one to each new base or I could calculate the nth term with ease.

Below you will see that one Triangle number always makes a Rhombus when combined with the previous Triangle number, or the next Triangle number. If you count the dots you will see a Square number is made.

```
                                            ■ □ □ □
                      ■ □ □                 ■ ■ □ □
      ■ □             ■ ■ □                 ■ ■ ■ □
  ■ ■                 ■ ■ ■                 ■ ■ ■ ■
```

If the pattern starts with one black dot, and I want to find how many black dots there will be on the fifth drawing I first find the total of white and black dots which will be $5^2 = 25$.

I then subtract one side i.e. 5 from the total 25 = 20

Then I divide that 20 by 2 to give the total white dots

Finally I subtract the white dots from the total dots:

25 − 10 = 15

This method can be put into an equation:

nth Triangle number = $n^2 - ((n^2 - n) / 2)$

Now I am going to find the nth polygon number.

If you look at the Polygon sequences, specifically at numbers under n, you can see that the number added along a row is a Triangle number, see below

n	
1	0
2	1
3	3
4	6
5	10
6	15
7	21
8	28
9	36
10	45

When n = 3, the number added each time is the 2nd Triangle number, when n is 4 you add the 3rd Triangle number, when n is 5 you add the 4th Triangle number each time and so on. To find the number added each time I have made the following equation:

$(n-1)^2 - (((n-1)^2 - (n-1))/2)$

Every sequence in a row starts with n, all I need to know now is how many times I add a specific Triangle number which I know to n.

If I want to know the 8th square number I add the 7th Triangle number to 8 two times.

8 + 28 + 28 = 64

If I want to know the 5th Pentagonal number I add the 4th Triangle number to 5 three times:

5 + 10 + 10 + 10 = 35

If I want to know the 10th Octagonal number I add the 9th Triangle number to 10, you guessed it, six times:

10 + 45 + 45 + 45 + 45 + 45 + 45 = 280

Just encase you didn't guess correctly, you will add a Triangle number 2 times less than the number of sides any polygon has.

The equation for the nth polygon number is thus:

n + (((n - 1)^2 – (((n-1)^2 – (n-1)) / 2)) x S - 2)

whereby:
n = a given number
S = sides of a Polygon

The only polygon number with a change between one difference and the next is one which is equal to the sequence 4, 10, 20, 34, 52… seen earlier are just the Hexagonal numbers i.e. 6 sided polygons.

1	3	4
6	4	10
15	5	20
28	6	34
45	7	52

To find the 6th number in the right hand sequence which if you recall is $(B^2 + A^2) \times 2)$ I can find the 6th number in the Hexagonal sequence shown on the left side and add 8. Basically I find the nth Hexagonal number and add $(2+ n)$.

Expressed as an equation this is:

$2 + 2n + (((n - 1)^2 - (((n-1)^2 - (n-1)) / 2)) \times 6 - 2)$

I make a grid of hexagonal numbers which can easily be converted to numbers in my other grid.

x	1	2	3	4	5	6	7	8	9	10	11	12
1	1	6	15	28	45	66	91	120	153	190	231	276
2	6	11	20	33	50	71	96	125	158	195	236	281
3	15	20	29	42	59	80	105	134	167	204	245	290
4	28	33	42	55	72	93	118	147	180	217	258	303
5	45	50	59	72	89	110	135	164	197	234	275	320
6	66	71	80	93	110	131	156	185	218	255	296	341
7	91	96	105	118	135	156	181	210	243	280	321	366
8	120	125	134	147	164	185	210	239	272	309	350	395
9	153	158	167	180	197	218	243	272	305	342	383	428
10	190	195	204	217	234	255	280	309	342	379	420	465
11	231	236	245	258	275	296	321	350	383	420	461	506
12	276	281	290	303	320	341	366	395	428	465	506	551

I can use the same method used before to find the number at any position. Let's look at 4 x 6 = 93 where A is 4 and B is 6.

To get 28 (the start of the sequence) I can do:

$(((A - 1)^2 - (((A-1)^2 - (A-1))/2)) \times 6 - 2)$

Now I must add one less numbers than B to the 28 i.e. 5 + 9 + 13 + 17 + 21 to get the answer 93.

The sequence always starts by adding 5 to the first number, and repeats in the same manner adding four more than the last time each time.

You can see how this pattern works:

5
5 + 4 = 9
5 + 4 + 4 = 13
5 + 4 + 4 + 4 = 17
5 + 4 + 4 + 4 + 4 = 21
5 + 4 + 4 + 4 + 4 + 4 = 25

So if I want to know the 6th number in the sequence I have shown above I do: 5 + (4 x (6 - 1)) = 25

Or in other words: 5 + (4 x (B - 1))

Both of these equations can be added together to get any number in the hexagonal grid, see below:

$7 + 2A + (((A - 1)^2 - (((A-1)^2 - (A-1))/2)) \times 6 - 2) + (4 \times (B - 1))$

To arrive at a number from the other grid you can do:

$2 + 2A + (((A - 1)^2 - (((A-1)^2 - (A-1))/2)) \times 6 - 2) = 34$

and then add $6 + (4 \times (B - 1))$

because the other patterns works like this:

6
$6 + 4 = 10$
$6 + 4 + 4 = 14$
$6 + 4 + 4 + 4 = 18$
$6 + 4 + 4 + 4 + 4 = 22$
$6 + 4 + 4 + 4 + 4 + 4 = 26$

To find the 6th number in the above sequence I can do
$6 + (4 \times (6 - 1)) = 26$

Or in other words: $6 + (4 \times (B - 1))$

Expressed as an equation this is:

$F + (C \times (F - 1)$

whereby
F = first number in sequences
C = common difference

Now I can replace $((B^2 + A^2) \times 2)$ with either:

$2 + 2A + (((A - 1)^2 - (((A-1)^2 - (A-1))/2)) \times 6 - 2) + 6 + (4 \times (B - 1))$

or

$8 + 2A + (((A - 1)^2 - (((A-1)^2 - (A-1))/2)) \times 6 - 2) + (4 \times (B - 1))$

I compare the difference between the multiplications and these polygon numbers to create the below grid.

x	1	2	3	4	5	6	7	8	9	10	11	12
1	0	0	0	0	0	0	0	0	0	0	0	0
2	-2	-1	0	1	2	3	4	5	6	7	8	9
3	-5	-2	1	4	7	10	13	16	19	22	25	28
4	-9	-3	3	9	15	21	27	33	39	45	51	57
5	-14	-4	6	16	26	36	46	56	66	76	86	96
6	-20	-5	10	25	40	55	70	85	100	115	130	145
7	-27	-6	15	36	57	78	99	120	141	162	183	204
8	-35	-7	21	49	77	105	133	161	189	217	245	273
9	-44	-8	28	64	100	136	172	208	244	280	316	352
10	-54	-9	36	81	126	171	216	261	306	351	396	441
11	-65	-10	45	100	155	210	265	320	375	430	485	540
12	-77	-11	55	121	187	253	319	385	451	517	583	649

Looking at the horizontal pattern it can be seen that numbers increase by a fixed Triangle number.

So if I wanted to know what 4 x 5 is I could find the 5th square number which is 36 and look at the grid to find which number to subtract 36 by.

36 – 16 = 20

And 20 is the answer to 4 x 5

Alternatively I could find the 4th pentagonal number which is 35 (1 isn't counted) and look at the grid to find the number to subtract this by.

35 − 15 = 20

To make this into an equation, I need to know how to find any of the numbers in the newest grid

The sequence I'm interested in goes like this:

0, -2, -5, -9, -14, -20, -27, -35, -44

For ease I'll turn these into positive values:

(B - 1) x 2 + 0.5 x (B -1) x (B - 2)

To turn them back into negative values I can do:

((B - 1) x 2 + 0.5 x (B - 1) x (B -2)) - (((B - 1) x 2 + 0.5 x (B -1) x (B - 2)) x 2)

The number of Triangle numbers I add to this number depends on the number of sides a polygon has, and these added numbers will be fixed.

If A = 3 I'll add the 2nd Triangle number each time
If A = 2 I'll add the 1st Triangle number each time
If A = 1 I'll add 0

The nth Triangle number = $n^2 - ((n^2 - n) / 2)$

but I need to add one less triangle

The following equation may be used:

$$((A-1)^2 - (((A-1)^2 - (A-1)) / 2))$$

To find any number in the grid I use the equation:

$$(((B - 1) \times 2 + 0.5 \times (B - 1) \times (B - 2)) - (((B - 1) \times 2 + 0.5 \times (B -1) \times (B - 2)) \times 2) + (A \times ((A-1)^2 - (((A-1)^2 - (A-1)) / 2)))$$

An equation for AB can then be made:

$$(((B + 1) + (B^2 - ((B^2 - B) / 2))) \times (A - 2)) - (((B - 1) \times 2 + 0.5 \times (B -1) \times (B - 2)) - (((B - 1) \times 2 + 0.5 \times (B -1) \times (B - 2)) \times 2) + (A \times ((A-1)^2 - (((A-1)^2 - (A-1)) / 2)))$$

Although this equation, and those like it, are utterly unpractical, I hope that you can appreciate AB more than you did before, and have greater confidence to unsimplify and understand Mathematics.

What equations would you like to see in AB?

Can AB ever be fully unsimplified?

www.ingramcontent.com/pod-product-compliance
Lightning Source LLC
Chambersburg PA
CBHW031941170526
45157CB00008B/3270